Acid Rain and Our Nation's Capital

A Guide to Effects on Buildings and Monuments

by Elaine McGee

Marble surfaces exposed to rain develop a rough "sugary" texture because the calcite grains are loosened as the edges dissolve in the rain water. Column capital volute, Jefferson Memorial, Washington, D.C.

For sale by the U.S. Government Printing Office
Superintendent of Documents, Mail Stop: SSOP, Washington, DC 20402-9328
ISBN 0-16-048068-X

A summer rain storm in Washington, D.C. (Memorial Continental Hall)

When polluted air mixes with rain, snow, and fog, acid precipitation forms. This acidity has caused people to worry about the environment; some reports show that acid rain has affected lakes, trees, and fish populations in the Northeastern United States and Canada. Another concern is its effect on historic buildings and monuments.

This booklet focuses on acid rain and its impact on our Nation's capital. Rain in Washington, D.C., has an average acidity of 4.2, about as acid as a carbonated drink and more than ten times as acid as clean, unpolluted rain. This booklet will define acid rain, explain what effects it has on marble and limestone buildings, and show, on a walking tour, some of the places in our Nation's capital where you can see the impact of acid precipitation.

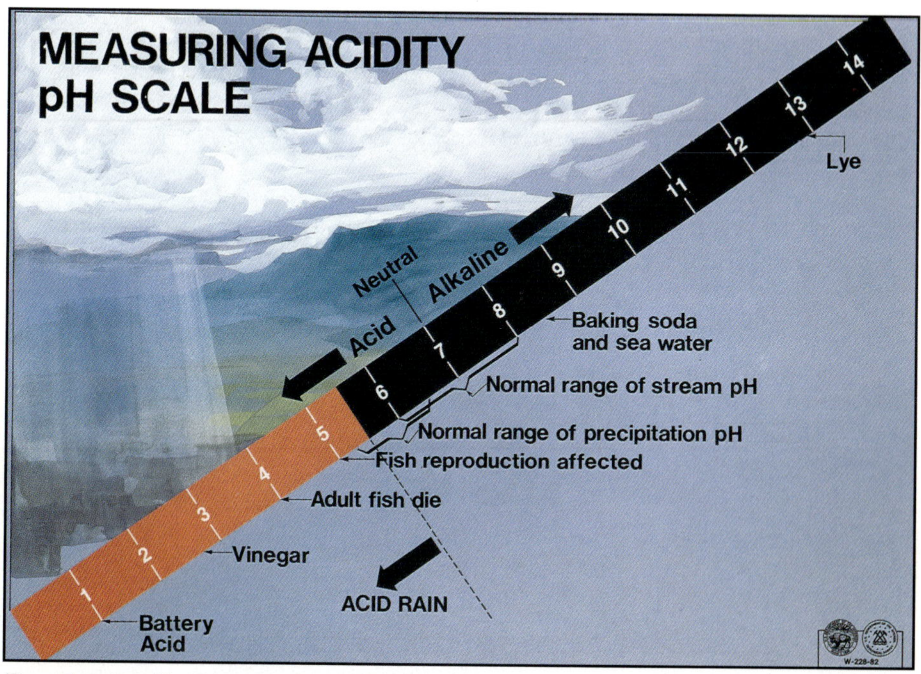

The pH scale: pH = 7 is neutral, neither acid or alkaline; smaller pH values are acid, larger pH values are alkaline. A liquid with a pH of 3 is ten times as acid as one with a pH of 4.

What is acid rain?

The term "acid rain" is commonly used to mean the deposition of acidic components in rain, snow, fog, dew, or dry particles. The more accurate term is "acid precipitation." Distilled water, which contains no carbon dioxide, has a neutral pH of 7. Liquids with a pH less than 7 are acid, and those with a pH greater than 7 are alkaline (or basic). "Clean" or unpolluted rain has a slightly acidic pH of 5.6, because carbon dioxide and water in the air react together to form carbonic acid, a weak acid. Around Washington, D.C., however, the average rain pH is between 4.2 and 4.4.

The extra acidity in rain comes from the reaction of air pollutants, primarily sulfur oxides and nitrogen oxides, with water in the air to form strong acids (like sulfuric and nitric acid). The main sources of these pollutants are vehicles and industrial and power-generating plants. In Washington, the main local sources are cars, trucks, and buses.

Acidity in rain is measured by collecting samples of rain and measuring its pH. To find the distribution of rain acidity, weather conditions are monitored and rain samples are collected at sites all over the country. The areas of greatest acidity (lowest pH values) are located in the Northeastern United States. This pattern of high acidity is caused by the large number of cities, the dense population, and the concentration of power and industrial plants in the Northeast. In addition, the prevailing wind direction brings storms and pollution to the Northeast from the Midwest, and dust from the soil and rocks in the Northeastern United States is less likely to neutralize acidity in the rain.

Wet and dry bucket collector, used to collect samples for measuring rainfall acidity.

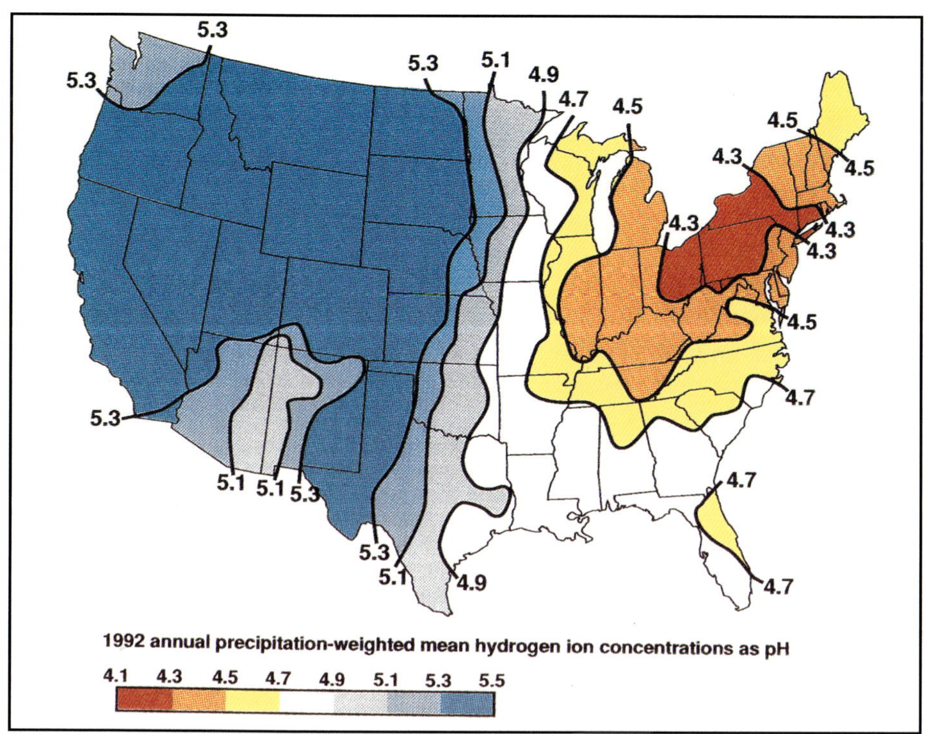

A pH distribution map shows areas in the continental United States of greatest acidity in the rain.

When you hear or read in the media about the effects of acid rain, you are usually told about the lakes, fish, and trees in New England and Canada. However, we are becoming aware of an additional concern: many of our historic buildings and monuments are located in the areas of highest acidity. In Europe, where buildings are much older and pollution levels have been ten times greater than in the United States, there is a growing awareness that pollution and acid rain are accelerating the deterioration of buildings and monuments.

Stone weathers (deteriorates) as part of the normal geologic cycle through natural chemical, physical, and biological processes when it is exposed to the environment. This weathering process, over hundreds of millions of years, turned the Appalachian Mountains from towering peaks as high as the Rockies to the rounded knobs we see today. Our concern is that air pollution, particularly in urban areas, may be accelerating the normal, natural rate of stone deterioration, so that we may prematurely lose buildings and sculptures of historic or cultural value.

What about buildings?

Many buildings and monuments are made of stone, and many buildings use stone for decorative trim. Granite is now the most widely used stone for buildings, monuments, and bridges. Limestone is the second most used building stone. It was widely used before Portland cement became available in the early 19th century because of its uniform color and texture and because it could be easily carved. Sandstone from local sources was commonly used in the Northeastern United States, especially before 1900. Nationwide, marble is used much less often than the other stone types, but it has been used for many buildings and monuments of historical significance. Because of their composition, some stones are more likely to be damaged by acidic deposition than others. Granite is primarily composed of silicate minerals, like feldspar and quartz, which are resistant to acid attack. Sandstone is also primarily composed of silica and is thus resistant. A few sandstones are less resistant because they contain a carbonate cement that dissolves readily in weak acid. Limestone and marble are primarily composed of the mineral calcite (calcium carbonate), which dissolves readily in weak acid; in fact, this characteristic is often used to identify the mineral calcite. Because buildings and monuments made of limestone and marble are more likely to be damaged by acid precipitation, they are the main focus of this booklet.

Memorial Bridge in Washington, D.C., is made of granite, the most widely used stone type.

Marble used as a trim on the First Bank in Philadelphia, Pennsylvania.

How do you recognize limestone and marble?

The main difference between limestone and marble is that limestone is a sedimentary rock, typically composed of calcium carbonate fossils, and marble is a metamorphic rock. Limestone forms when shells, sand, and mud are deposited at the bottom of oceans and lakes and over time solidify into rock. Marble forms when sedimentary limestone is heated and squeezed by natural rock-forming processes so that the grains recrystallize. If you look closely at a limestone, you can usually see fossil fragments (for example, bits of shell) held together by a calcite matrix. Limestone is more porous than marble, because there are small openings between the fossil fragments. Marble is usually light colored and is composed of crystals of calcite locked together like pieces of a jigsaw puzzle. Marble may contain colored streaks that are inclusions of non-calcite minerals.

Limestone is made of fossil fragments, held together with calcite; the shell near the center is about 1 cm across. Botanic Gardens building Washington, D.C.

Marble is made of calcite crystals (white) and some colored grains of mica inclusions; the grains in a marble are locked together like jigsaw puzzle pieces.

How does acid precipitation affect marble and limestone buildings?

Acid precipitation affects stone primarily in two ways: *dissolution* and *alteration*. When sulfurous, sulfuric, and nitric acids in polluted air react with the calcite in marble and limestone, the calcite dissolves. In exposed areas of buildings and statues, we see roughened surfaces, removal of material, and loss of carved details. Stone surface material may be lost all over or only in spots that are more reactive.

You might expect that sheltered areas of stone buildings and monuments would not be affected by acid precipitation. However, sheltered areas on limestone and marble buildings and monuments show blackened crusts that have spalled (peeled) off in some places, revealing crumbling stone beneath. This black crust is primarily composed of gypsum, a mineral that forms from the reaction between calcite, water, and sulfuric acid. Gypsum is soluble in water; although it can form anywhere on carbonate stone surfaces that are exposed to sulfur dioxide gas (SO_2), it is usually washed away. It remains only on protected surfaces that are not directly washed by the rain. Gypsum is white, but the crystals form networks that trap particles of dirt and pollutants, so the crust looks black. Eventually the black crusts blister and spall off, revealing crumbling stone.

When marble is exposed to acidic rain, sharp edges and carving details gradually become rounded. Antefixes, roof of the Philadelphia Merchants' Exchange (built in 1832).

Blackened crusts on sheltered portions of the limestone Chicago Tribune Building, Chicago, Illinois.

Formed as a result of air pollution, gypsum alteration crusts have blackened, blistered, and spalled from a marble baluster at the Organization of American States building, Washington, D.C.

Scanning electron microscope photograph of gypsum crystals with dirt and pollution particles trapped by the network of crystals. The scale bar is 10 micrometers long.

A marble column at the Merchants' Exchange in Philadelphia shows loss of material where the stone is exposed to rain and blackening of the stone surface where the stone is sheltered from rain.

Where can we see the effects of acid precipitation?

Washington's buildings and monuments use many different stone types. Marble and limestone buildings are the most likely to show damage, because they are more affected by acidic precipitation and urban pollution. As you follow the tour described in this book, see how granite and sandstone buildings compare with the marble and limestone in the same environment.

This guide will help you recognize some geologic features of buildings, in addition to their historical and architectural aspects, wherever you travel. However, remember one important point when examining buildings and monuments for deterioration: stone deterioration has many causes. Although acid precipitation and urban pollution can accelerate stone deterioration, people, pigeons, and other organisms may also harm our stone structures. In addition, the process of weathering has been going on since the Earth first had an atmosphere. Although we can observe deterioration of the stone, it is hard to determine how much of the deterioration is from acid precipitation and how much is from other causes.

Pigeons sitting on the statue heads have created distinctive deterioration on this building.

Flowers and grasses have grown in the cracks between stones on this church.

This limestone column in the Lincoln Memorial is darkened and dirty from people's hands touching the stone.

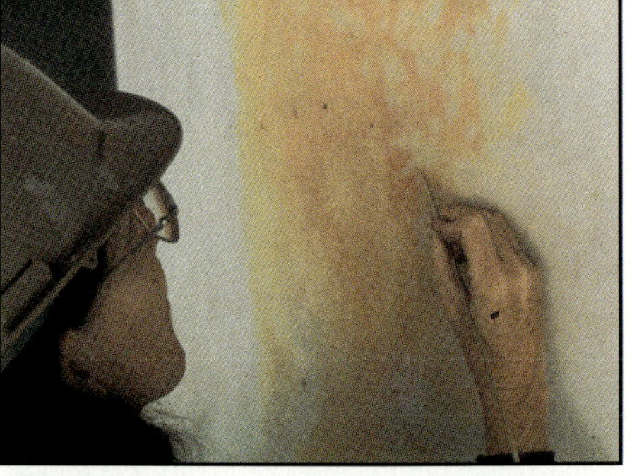

Microorganisms have caused this stain to appear on a marble column at the Jefferson Memorial.

What are we doing about acid rain?

Scientists from many disciplines are studying acid precipitation and its impact. The National Acid Precipitation Assessment Program (NAPAP), a Federal program involving representatives from more than a dozen Federal agencies, has sponsored studies on how acid rain forms and how it affects lakes, crops, forests, and materials. Because buildings and monuments cannot adapt to changes in the environment, as plants and animals can, historic structures may be particularly affected by acid precipitation. Scientists are studying effective control technologies to limit the emissions from power plants and automobiles that cause acid rain. The impact and usefulness of regulations that would require limits on air pollution are also being studied. Finally, scientists are examining the processes of deterioration to find effective ways to protect and repair our historic buildings and monuments. Agencies like the National Park Service, which are charged with protecting and preserving our national heritage, are particularly concerned not only about the impact of acid rain but also about making the best choices for maintaining and preserving our historic buildings and monuments.

Beginning in 1984, the National Acid Precipitation Assessment Program sponsored exposure site studies of limestone and marble, to examine the contribution to stone deterioration that comes from acid precipitation.

A field guide to buildings in our Nation's capital

Washington, D.C., has many buildings of historic and cultural significance, and many of them are made of marble and limestone. This self-guided tour will point out damage to buildings and monuments in our Nation's capital that may have been caused by acid precipitation. Similar effects may be found in other cities as well.

Places to visit have been divided into several areas, so the trip can be done either in segments or all in one day. A suggested tour route is described within each area. A car provides the most efficient transport between areas, but parking may be hard to find. The Metro subway system can easily be used to visit all areas except the Jefferson and Lincoln Memorials. The closest Metro stations in each area are shown on the map. You will need comfortable walking shoes, and you may want to bring along a camera, a hand lens (about 10x magnification) for observing details of minerals and weathering, and a pair of binoculars for closer examination of inaccessible areas.

The area around the National Capitol

This area includes the Capitol building, the Peace Monument, the Grant Memorial, and the Botanic Gardens. We begin the tour at the southeast corner of the Capitol, and go clockwise around the Capitol (along the south, west, and then north sides). We then follow a walkway heading west, from the northeast corner of the Capitol, to see the Peace Monument at the intersection of First Street and Pennsylvania Avenue, NW. We continue south along First Street to the Grant Memorial and then south again to the Botanic Gardens. Total distance is about one kilometer, or about three-quarters of a mile.

The Capitol Building—Site 1

The Capitol was built in stages; the cornerstone of the main building was laid in 1793, the north wing was completed in 1800, and the south wing was completed in 1807. Both wings were burned by the British in 1814. The capitol was then rebuilt, and it has been modified several times throughout the years. A major program of cleaning, replacement, and repair was begun in the late 1980s. The center building of the Capitol is painted sandstone, but the north and south wings, housing the Senate and the House chambers, are marble. Around the Capitol we will observe various examples of dissolution and blackened alteration, especially on the marble balustrade that surrounds the south, west, and north sides of the building.

Beginning at the southeast corner of the building, by using binoculars we can see some areas of blackened alteration in the Corinthian column capitals. A more accessible example is found under the overhang of the large square ends of the marble

The United States Capitol building.

balustrade at the southeast corner of the building. The black crust is made of gypsum plus dirt that accumulates in sheltered areas. No black crust is present along the cracks between the stones; rain water probably flows in these areas, dissolving the gypsum and preventing accumulation of a crust. Not all black areas on this baluster are gypsum; in some places near the bushes, you can see greenish-black moss growing on the stone. The top surfaces of the marble balustrade are coarse and rough, because of dissolution between grains, compared to areas that are protected from running or washing water where the black alteration crust forms.

Blackened alteration has accumulated under the overhanging edge of this marble balustrade corner. Southeast corner, U.S. Capitol Building.

Pock marks in marble columns, south side of U.S. Capitol building. Silicate mineral inclusions in the marble loosen and fall out when the calcite around them is dissolved by acid rain.

Another dissolution feature of marble is the pock-mark effect on the square bases of the building columns. Silicate mineral inclusions in the marble were loosened by the dissolution of the surrounding calcite, causing the inclusions to fall out of the stone. A particularly good example of this is found on the fourth column west from the southeast corner of the Capitol building. The pock-mark dissolution is also found at several other places on the building.

We will follow the marble balustrade around the building, noting differences in deterioration. Some parts of the balustrade have obviously been replaced, thus enabling us to observe various stages in the stone deterioration. The edges of the balusters are sharp when new and become rounded as they age. Blackened alteration crusts have accumulated on the sheltered sides of the balusters and under the overhanging top of the balustrade. In

The marble balustrade on the west side of the Capitol building shows both dissolution and alteration.

A recently replaced marble baluster at the Capitol has edges that are only beginning to round.

Black alteration crusts under the west balustrade of the Capitol have begun to spall (peel) off, revealing crumbling white marble underneath.

some spots under the rail the blackened crust has spalled off, exposing fresh surfaces and more vulnerable stone. Some carvings on the balustrade corners are worn, whereas others have blackened alteration; this difference in weathering may be due to local effects of wind and rain. Along the steps leading to the terrace on the west side of the Capitol, gypsum has accumulated on large areas of the wall. Gypsum can accumulate on any surface that is not washed by water.

As you walk north along the west side of the Capitol, look at the central part of the building. The walls here are painted sandstone. Despite recent restoration of the building, you can see evidence of past stone deterioration, including the accentuated lines from bedding in the stone and the pock marks where rounded inclusions have disappeared. We will see an example of this same sandstone that is not painted in the buildings near 17th Street.

At the northeast corner of the Capitol building, the marble balustrade ends in square blocks like the ones we first examined. Here you can see an example of preferential dissolution where the silicate mineral inclusions remain and the calcite around them has been dissolved away. Also, on the north side of this block, examine the blackened grains on the top surface with a hand lens. Not all of the black material you see on stone is gypsum; some is of biological origin, probably algae or a fungus.

A marble block that forms the northeast corner of the Capitol balustrade shows preferential erosion of the calcite around a silicate mineral inclusion.

To continue the tour, follow the pathway that heads west along the north side of the Capitol, towards First Street. As you approach First Street you will see a sandstone, diabase (a dark igneous rock), and granite fence with various carvings. Because these stone types are resistant to acid attack, the carvings show little damage.

The Peace Monument—Site 2

The Peace Monument, dedicated in 1878, is made of Italian marble.

This monument, dedicated in 1878, is made of marble from Carrara, Italy. The statue does not show much damage, but if you look closely, you can see alteration crusts (some are light orange) in protected places and graininess and roughness in places that are exposed to rain. Carved statues present varied surfaces that direct rain washing and runoff.

Sites on the tour:

1. THE CAPITOL BUILDING
2. THE PEACE MONUMENT
3. THE GRANT MEMORIAL
4. BOTANIC GARDENS BUILDING
5. JEFFERSON MEMORIAL
6. LINCOLN MEMORIAL
7. CAPITOL GATEHOUSE

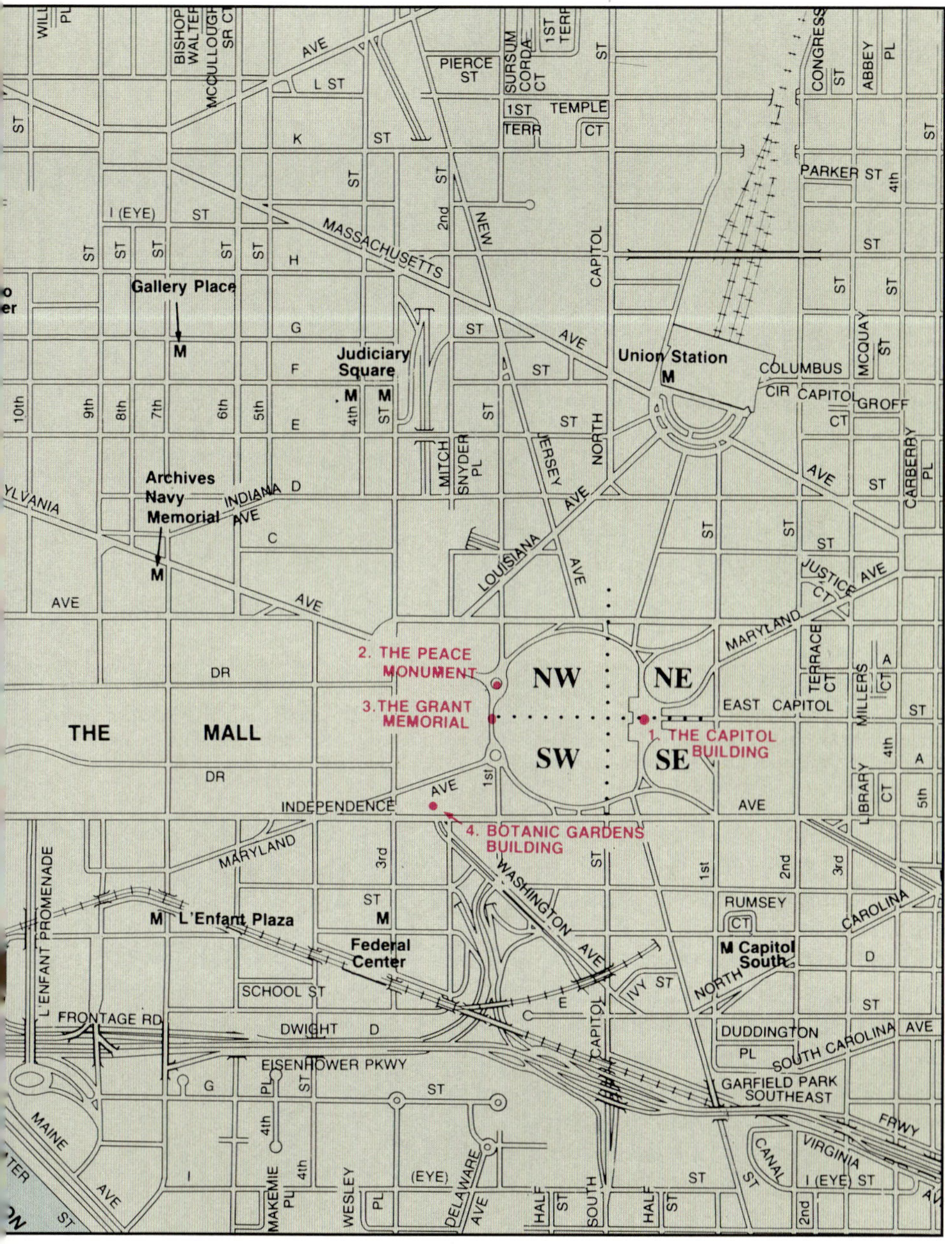

8. ORGANIZATION OF AMERICAN STATES BUILDING

9. DAR—CONSTITUTION HALL

10. DAR—MEMORIAL CONTINENTAL HALL

11. CORCORAN BUILDING

12. RENWICK GALLERY

13. FEDERAL TRIANGLE BUILDINGS

14. WASHINGTON MONUMENT

Continue south along First Street to the Grant Memorial on your right.

The Grant Memorial—Site 3

The Grant Memorial consists of bronze statues on marble bases; although the statues were cleaned, the staining on the marble remains.

This memorial, dedicated in 1922, consists of a group of bronze sculptures mounted on marble bases. Bronze weathers outdoors if it is not cleaned and waxed regularly. Like stone, bronze dissolves where it is exposed to rainfall, developing a green color and a pitted surface, and it also alters in sheltered areas, with accumulation of a blackened layer. The most notable stone deterioration visible here is the green stain on the marble bases, caused by runoff from the weathered bronze. The green stain does not damage the marble, but it is unattractive, and there are no methods currently available to remove the stain without damaging the marble.

The light-green color on the bronze statues at the Grant Memorial is typical of the way bronze weathers when it is exposed to acidic rain; the rain dissolves some of the metal and causes staining of the white marble bases.

Continue south, crossing Maryland Avenue, to the Botanic Gardens.

Botanic Gardens Building—Site 4

The Botanic Gardens building is made of limestone.

This limestone building was built in 1931. Like many of the limestone buildings in Washington, this building has been cleaned, so it does not have an accumulation of surface dirt. However, the cleaning and regular washing by rainfall have accentuated the fossils in the stone, which dissolve less readily than the calcite matrix. Some of the sculpted heads above the arches of the building show small black crusts. On the east side of the building, microorganisms naturally present in the stone contribute to the deterioration (blackening) of the stone, where water drips from a joint in the roof.

Fossil details stand out on the limestone at the Botanic Gardens building; the fossils are more resistant to dissolution than the calcite matrix that holds the fragments together.

The Botanic Gardens Building is the last stop in the Capitol area of the tour; you may wish, however, to see some bronze alteration on the Garfield Memorial (First St. and Maryland Ave.) and the accentuated fossils with surrounding algae or fungi on the limestone posts near the Capitol Reflecting Pool. The next stop, the Jefferson Memorial, is about 3.5 kilometers (2.2 miles) from the Capitol.

Jefferson Memorial—Site 5

The Jefferson Memorial is made of marble and was dedicated in 1943.

The Jefferson Memorial is a marble building, dedicated in 1943. One of the most striking deterioration features to observe here is the loss of silicate mineral inclusions in the marble columns because of dissolution of the calcite matrix. Close examination of the grooves shows flakes of mica and sometimes grains of pyrite. Blackened crusts are visible on the column capitals that are sheltered from rain and from regular washing of the monument.

The National Park Service began a survey of the condition of this

Several of the column shafts at the Jefferson Memorial have grooves that follow the inclusion traces in the marble, where the mineral inclusions have weathered out and been lost.

A close look at some of the weathered grooves in the columns shows that small bits of mica and pyrite remain.

memorial and the Lincoln Memorial in 1992. The results will be used to help make decisions on treatment, cleaning, and preservation. The information gathered from the survey will serve as a known baseline for the condition of the stone, so that future changes in the condition of the buildings can be assessed. In May 1990, a part of one of the column capitals (called a volute) broke off and fell onto the northwest portico. This failure raised concern about all the volutes at the Memorial, so several other cracked volutes were removed, and studies are being conducted to determine why they cracked. Because of where and how the volute broke, it is unlikely that acid rain or air pollution contributed to the failure. The broken pieces will probably be replaced, but only when the reason for their failure is understood, so that an appropriate replacement technique can be chosen.

Part of one of the column capitals at the Jefferson Memorial broke off and fell onto the portico in 1990.

__The next stop is the Lincoln Memorial, 1.8 kilometers (a little more than a mile) northeast of the Jefferson Memorial.__

Lincoln Memorial—Site 6

The Lincoln Memorial, dedicated in 1922, is made of marble from Colorado. This building has few alteration crusts, in part because it receives regular cleaning and in part because of the design of the building. Except for the features around the entablature (the edge of the roof), there are few sheltered areas where alteration crusts can accumulate. With the aid of binoculars, you can see some alteration crusts along the underside of the roof overhang; in these places the marble is very badly crumbled under the alteration crusts. Some columns show preferential weathering or loss of inclusions, but some of this damage might be from graffiti removal. The most visible dissolution feature is sugaring, where the stone has lost its polish and the surface now feels rough. Visitors have affected this popular memorial too; several of the columns, especially the limestone columns inside the chamber, show darkening and rounding of edges where visitors have touched them over the years.

One interesting feature at the Lincoln Memorial is differences in stone condition that must come from variations in the stone. At several places around the outside of the memorial, adjacent blocks of marble show very different surface roughness. Since the blocks of stone have the same orientation with respect to wind, rain, and pollution, the difference in condition cannot be due to exposure and must be related to basic characteristics in the stone that was used.

The Lincoln Memorial is made of marble and was dedicated in 1922.

Some columns at the Lincoln Memorial have flattened chalky areas where inclusions have weathered differently from the surrounding calcite.

Under the roof overhang is one of the few places at the Lincoln Memorial where alteration crusts have developed on the marble.

The marble guttae on the roof overhang are crumbling and falling apart underneath blackened alteration crusts.

Buildings along 17th Street, NW

Some notable examples of stone deterioration are included in this part of our tour. We will begin this segment at the northeast corner of 17th Street and Constitution Avenue, with the Capitol Gatehouse. We will then cross 17th Street and examine several buildings along 17th Street as we walk north towards Pennsylvania Avenue. The total distance is about three-fourths of a kilometer (half a mile).

Capital Gatehouse—Site 7

The Capitol Gatehouse, now located at 17th Street and Constitution Avenue, is made of the same sandstone used in the White House and the center part of the Capitol, but it was left unpainted. Deterioration of this stone is due to the clay it contains, not to the effects of acid rain.

This small sandstone building was built around 1828 at the west entrance to the Capitol. In 1880 it was moved (along with a twin and four gateposts) to its present site. This building is made of the same sandstone that was used in the central part of the Capitol and in the White House. Three types of deterioration are readily visible at the gatehouse: spalling, pock marks, and preferential weathering of clay layers in the stone. This stone may be more degraded than stone in the Capitol or the White House, because of variations in stone quality and maintenance to the buildings and because it has never been painted.

This kind of sandstone was soon found to be a poor building stone because of its tendency to spall. (detail on Capitol Gatehouse)

To continue, we will cross 17th Street and examine parts of several buildings as we walk north.

Organization of American States Building—Site 8

The Organization of American States Building is made of marble and was dedicated in 1910.

Blackened gypsum crusts may blister and spall off, exposing a crumbling stone surface to further pollution.

This marble building was dedicated in 1910. Two sculptures in the front of the building show some alteration crusts in sheltered areas and dissolution in exposed areas. In back of the building the marble balusters on the patio are covered with blackened crusts, especially on the sides facing the garden. In many places the crusts have blistered or spalled off, exposing new surfaces to alteration. In general, the patio sides of the balusters are in much better condition than the sides that face the garden, perhaps because washing of the patio has washed off the gypsum crusts on that side of the balusters.

DAR—Constitution Hall—Site 9

DAR—Constitution Hall is made of limestone and was built in the 1930's.

Some of the blackening on limestone surfaces may be from algae or fungi that readily grow in the rough surface in Washington's warm, humid climate.

The main damage on this limestone building, built in the 1930's, is the blackening of the side balustrade from algae or fungi. The stone is porous and therefore retains moisture, thus encouraging growth of organisms. The limestone in this building is quite uniform and shows little preferential dissolution, except in a few places. On the top of the balustrade along C Street, for example, some of the calcite matrix has dissolved from around the fossil fragments, and some holes are filled with calcite crystals.

DAR—Memorial Continental Hall—Site 10

Memorial Continental Hall, built in 1909, is part of the Daughters of the American Revolution building complex.

The porch area on the south side of this marble building built in 1909 is a good place to look at some contrasts in marble deterioration. Parts of the balustrade have been replaced, as shown by differences in color and surface roughness of the stone. The exposed stone surface

Carvings at the base of the columns on the south side of Memorial Continental Hall show that carved details and sharp edges remain on sheltered areas.

along the top of the balustrade is rougher than the surfaces in more sheltered areas. The columns on this porch are carved around the base, so you can examine the effects of exposure to rain on the carving details. The more exposed carvings have lost their sharp edges and definition compared to the sheltered carvings. The bases of the columns contain small amounts of pyrite, which is more resistant to weathering than is the calcite in the marble surrounding the pyrite. The sheltered part of the windowsill support on the west side of the porch shows an alteration crust, a dull gray accumulation on the stone surface.

Pyrite grains stand in relief where calcite and micas have weathered out of the marble at Memorial Continental Hall.

On an exposed portion of the carving on the columns at Memorial Continental Hall, the edges of the marble have rounded and the surface has roughened.

A dull gray surface on the marble on the window sill support shows where an alteration crust is just beginning to develop.

At the corner of 17th and C Streets, turn left and walk north along 17th Street. On our way to the Corcoran Gallery, we will pass the Red Cross building (marble, 1917). Some of the same types of marble deterioration observed at other locations are also present here.

Corcoran Gallery—Site 11

The Corcoran Gallery is built mostly of marble.

The Corcoran Gallery is marble with a granite base. It was built in 1879 and enlarged in 1927. Ornate carvings around the roof, doors, and windows have blackened crusts of gypsum, as do parts of the marble pedestals supporting the bronze lions at the front door. The marble bases also have inclusions that stand out above the surrounding calcite, which has been dissolved away.

Marble bases for bronze lions outside the entrance to the Corcoran Gallery have feldspar inclusions that stand in relief compared to the roughened surrounding calcite.

31

Continuing north along 17th Street towards Pennsylvania Avenue, you will see several modern granite office buildings and the Executive Office building (formerly the State-War and Navy building), which was built from granite and completed in 1888. These granite buildings show little deterioration. Turn right onto Pennsylvania Avenue and proceed to the Renwick Galley on the northeast corner of the intersection of 17th Streets and Pennsylvania Avenue.

Renwick Gallery—Site 12

The Renwick Gallery, made of brick and sandstone, was completed in 1859.

This building of brick and sandstone, completed in 1859, is interesting from a stone preservation point of view. The decorative sandstone panels were badly deteriorated, so in 1968 the panels were saturated with epoxy to strengthen them. This treatment actually accelerated the deterioration because when water penetrated behind the epoxy-filled area, large portions of the treated panels spalled off. A second renovation attempt was therefore necessary two years after the first, and the present panels are cast sandstone. A post of the original sandstone stands at the southeast corner of the building.

Casts of ground sandstone and epoxy replaced the original carved sandstone decorative trim at the Renwick Gallery when a first attempt to preserve the carved stone failed.

The next part of the tour begins at 15th Street and Pennsylvania Avenue S. To get there, walk east along Pennsylvania Avenue, past Blair House and between Lafayette Park and the White House. Lafayette Park has a number of bronze statues that have been cleaned fairly recently. The White House is built of sandstone that was painted white; the paint was used in part to improve the durability of the stone. After you pass the White House, you will come to the Treasury Building. Turn right onto 15th Street and walk south, towards the Washington Monument and the Mall. The total distance from the Renwick to the corner of 15th and Pennsylvania is about three-fourths of a kilometer (half a mile).

Federal Triangle Buildings—Site 13

All of the Federal buildings that form the Federal Triangle (between Pennsylvania Avenue, Constitution Avenue, and 15th Street) are made of limestone.

On the east side of 15th Street, beginning at E Street, is the Commerce Department building, which was constructed of limestone in the 1930's. This building is part of the Federal Triangle, a cluster of Federal office buildings in the area bounded by Pennsylvania Avenue, Constitution Avenue, and 15th Street, built primarily during the New Deal administration of President Franklin D. Roosevelt. Some sculptures on the buildings were done by participants in the WPA program. These buildings were cleaned in the 1960's, probably by sandblasting. Look for fossils in relief and alteration crusts in some sheltered places on the carved work. Some of the blackening on this building is from dirt and organic material trapped or growing in the rough surface of the stone.

Continue south on 15th Street to Constitution Avenue. From the corner of 15th Street and Constitution, follow some of the foot paths half a kilometer (three tenths of a mile) to the Washington Monument.

Washington Monument—Site 14

This monument was begun in 1848, but it was not finished until 1885; the change in color about 150 feet up marks a change in the type of marble used to face the monument. Although it is made of marble, its smooth, straight shape and the massive blocks used in this monument have minimized the effect of acid precipitation. Dissolution does occur in a few areas, but the amount of stone material lost from dissolution is insignificant compared to the mass of the stone.

The straight shape and massive stones in the Washington Monument minimize the impact of acid precipitation to this important landmark.

Our tour ends here, but there are many more stone buildings and monuments in Washington and in other cities that may also show the effects of urban pollution and acid precipitation. However, as seen on this tour, not all of the deterioration of stone buildings is caused by acid rain. To protect our historic stone buildings and monuments, we need to limit air pollution that contributes to acid rain deterioration, and we need to develop effective maintenance and preservation procedures that will not further harm the stone.

Glossary of Geologic and Architectural Terms

Antefix - an ornament along the edge of a roof, often shaped like a shell or shield

Baluster - a post or support for a handrail

Balustrade - a rail and row of supporting posts, especially along a stair or porch

Bronze - a metal alloy of copper and tin; although brown when fresh, it weathers to a more commonly seen green color

Calcite - a mineral composed of calcium carbonate: $CaCO_3$

Column Capital - The top portion of a column. There are three main types; from least ornate to most ornate, they are Doric, Ionic, and Corinthian

Diabase - a dark-gray to black, fine-textured igneous rock, composed of the minerals feldspar and pyroxene

Feldspar - a common mineral composed of silica (Si), aluminum (Al), and calcium (Ca), potassium (K), or sodium (Na): $CaAl_2Si_2O_8$, $KAlSi_3O_8$, $NaAlSi_3O_8$

Fossil - remains or trace of a plant or animal preserved in a rock from some past, prehistoric time

Granite - a light or speckled, coarse-grained igneous rock composed of quartz, feldspar, and mica

Guttae - small, drop-like rounded ornaments under a roof overhang

Gypsum - a mineral composed of hydrated calcium sulfate: $CaSO_4 \cdot 2H_2O$

Igneous - formed from melted or partially molten rock

Inclusion - a mineral phase that is distinct (in composition or appearance) from the main mineral components of a rock

Limestone - a sedimentary rock, chiefly composed of calcium carbonate; formed from the shells of marine animals

Marble - a metamorphic rock, chiefly composed of calcium carbonate

Metamorphic - changed; usually by heat or pressure to a different form but the same composition

Mica - a silicate mineral that forms sheets or layers and contains aluminum, hydroxyl, and alkali ions: $K_2Al_4Si_6Al_2O_{20}(OH,F)_4$, $K_2(Mg,Fe)_6Si_6Al_2O_{20}(OH,F)_4$

Mineral - a naturally occurring inorganic substance with an ordered structure; rocks are made of one or more minerals

Portico - a porch, with a roof supported by columns

Pyrite - a mineral ("fool's gold") composed of iron sulfide: FeS_2

Pyroxene - a silicate mineral containing two metal oxides: $CaMgSi_2O_6$, $CaFeSi_2O_6$, $(Mg,Fe)SiO_3$

Quartz - a mineral composed of silicon dioxide, or silica: SiO_2

Sandstone - a sedimentary rock composed chiefly of quartz grains

Sedimentary - formed from particles that were transported by water or wind and deposited in layers

Volute - a scroll-like ornament on an Ionic column capital

Further Reading

Acid rain: The facts, 1988: available from Inquiry Centre, Environment Canada, Ottawa, Ontario, Canada K1A 0H3.

Amoroso, G.G., and Fassina, V., 1983, Stone decay and conservation: New York, Elsevier, 453 p.

Hannibal, J.T., and Park, L.E., 1992, A guide to selected sources of information on stone used for buildings, monuments, and works of art: Journal of Geological Education, v. 40, p. 12-24.

Pickering, R.J., 1987, Acid rain: U.S. Geological Survey Open-File Report 87-399.

U.S. Geological Survey, Building stones of our Nation's Capital: Reston, Va., U.S. Geological Survey general-interest publication.

Winkler, E. M., 1973, Stone: Properties, durability in man's environment: New York, Springer-Verlag, 230 p.

Limestone

Sandstone

Marble

Granite

Photographs by the author · Edited by Kathleen Gohn · Publication design by Elizabeth Junek